P9-DGU-677

AMERICAN BARNS

A PICTORIAL HISTORY

Jill Caravan

AN IMPRINT OF
RUNNING PRESS BOOK PUBLISHERS
Philadelphia • London

9 8 7 6 5 4 3 2 1
Digit on the right indicates the number of this printing

ISBN 1-56138-471-2

Library of Congress Cataloging-in-Publication Number 93-87595

This book was designed and produced by
Todtri Productions Limited
P.O. Box 20058
New York, NY 10023-1482

Producer: Robert M. Tod
Book Designer: Mark Weinberg
Photo Editor: Edward Douglas
Editors: Mary Forsell, Don Kennison
Production Coordinator: Heather Weigel
DTP Associate: Jackie Skroczky
Typesetting: Mark Weinberg Design, NYC

Printed and Bound in Singapore by Tien Wah Press

Published by Courage Books
An imprint of Running Press Book Publishers
125 South Twenty-second Street
Philadelphia, Pennsylvania 19103-4399

CONTENTS

INTRODUCTION

Near my home in Lehigh County, southeastern Pennsylvania, there stand the stony remains of a barn, just a corner of the building's former foundation. It keeps vigil at the entrance to an office complex, left there by the developer as a landmark of sorts.

The developer probably intended it as a monument to the family farm—complete with barn, outbuildings, and farmhouse—that once occupied those lands before crews tore them down to make room for the brick slabs of leasable space that now sit there. However, in that same decision, that developer made a statement on the passing of an important part of America.

The family farm, with the barn as its most notable architectural feature, is disappearing quickly. Signs that read "Coming Soon..." and "Lots For Sale" have

RIGHT: Commanding a central position in the landscape, the barn has always been a primary source of pride to the farmer. Oftentimes, the date of construction is prominently displayed on the building.

OPPOSITE: With the coming of spring, the focus of the farm turns to the outdoors. In the fall, the barn will be all-important, providing storage for crops and machinery and shelter for animals.

Those who are lucky enough to have grown up in a rural setting will always have special images of the barn—such as this one surrounded by snow at dawn—in their memories.

become the new landmarks of rural America, fast on its way to becoming suburbia.

Although the same trend is running its course through most of the industrialized Western nations, the United States gives us the clearest picture of the vanishing family farm. From 1950 to 1990 the number of farms in the country plummeted from more than five million to just two million. Likewise, the number of farm families shrank from 23 million to less than six million. Over the same period farm production more than doubled, and farm size climbed from 215 to 450 acres.

What these converging trends have given us is a farm economy in which the largest twenty percent of American farms produce nearly eighty percent of the farm output. The smallest fifty percent—mostly family farms—give us just ten percent of our crops and animals.

Today's family farm, if successful, is a specialized business, relying on regular infusions of large amounts of money and energy. It's a one- or two-crop operation—maybe with a general garden plot for production of more diverse fare for the farm family's table.

Little remains of the ideal, pastoral setting that comes to mind for much of the nonfarming segment of the population at the mention of the word "farm." In no aspect of the farm environment is this more evident than in the increasing dearth of that dominant farmland structure, the barn. Large, corporate farms need many fewer barns and generally prefer to install large, central, prefab, aluminum structures that may be quite utilitarian but bear little resemblance to what we have come to know as a barn.

If I wax a bit on the critical side concerning those forces that would remove even one more barn from our landscape, please chalk it up to the sweeter-with-time memories from my child-

Many barns built several decades ago now sport modern additions, such as the very nontraditional barn doors on this structure. Rather than breaking with tradition, such contemporary features simply carry on the barn-building custom of incorporating whatever worked best.

hood. My preadult years were spent a little less than a two-hour's drive northwest of the more developed region where I now reside, in a small town in Pennsylvania's hard-coal region. Here, the mines and woodlands defined the mountainsides, while family farms dominated the valleys between. A very traditional, three-story barn (of the Pennsylvania standard type, which will be examined later) stood nearby my bedroom window.

That friendly old structure is part of many a memory from my formative years. I played with the kids of the farm family among the bales of hay in the loft. We staged many an Old West shoot-out through the strange and wonderful doors and shoots. We learned about animals from the cows and horses housed in the lower level. And, always, that uniquely barnlike smell—a mix of hay, animals, and machinery—hung in the air.

Spring on the farm as a whole was the true beginning of the year, marked by working the fields, planting, and new growth. The barn became less interesting at this time of the year. The hay, in which we built our forts and tunnels, was largely used up. The animals spent most of their time grazing out in the pasture. The massive machinery was at work in the fields. And the warmer days beckoned us to new pursuits.

During the hottest days of summer we would sometimes find a bit of relief in the barn, especially in the dark recesses next to the lower level of the stone foundation. Regardless of how hot or how cold it was outside, those nooks and crannies always offered us a more stable temperature. In the summer they were cool, and in the winter they were relatively warm.

It was in autumn when the barn really came into its own. The hayloft was beginning to fill. The animals sought refuge from the colder

FOLLOWING PAGE:
The traditional symbol of the rural Vermont countryside—and the one that attracts so many visitors—is the barn.

mornings. All that wonderful, imagination-stirring machinery was returning.

In winter the barn was cool enough for some pretty strenuous play, yet warm enough that a winter coat might be shed on occasion. The animals stayed pretty close to their shelter and ready food supply.

But that's "my" barn, the one that lives in my memory. You, coming from a different region of the country and from a different life—perhaps one far removed from any direct experience in a barn—probably think of something quite different when you call up your personal vision.

In more modern times barns have provided huge canvases for artists wanting to make visual statements on a grand scale, as with this patriotic design.

OPPOSITE: Much to the delight of passing motorists, many farmers take the time during the Yuletide to create lighted Christmas decorations on a grand scale, using their barns as backdrops.

CHAPTER ONE

BASIC STYLES

There have been many different styles of barns. Unlike most architectural styles, however, the differing barn styles did not arise as a result of rebellion with former styles. To the contrary, builders of barns were driven entirely by the need for full function within and around the structure they were erecting. If something from a former style fit into their plans and needs, they incorporated it into their barns. If something new was needed, they adapted it into a new style to that end.

As a result, different barn styles for different needs arose in particular settings, but always with a similar basic design. It's been that way through most of the history of America.

ORIGINS

The very earliest British and European settlers to the North American continent brought with them the same barn styles they were acquainted with in the Old Country. Since they came from a relatively small area of the world, their building styles were similar. However, there were significant differences.

The English in early Jamestown (1607) and Plymouth (1620s) employed the frame-building style, with thatch and clapboard covering, for their new homesteads, houses and barns alike. Hewing such structures from the wilderness materials and using makeshift building techniques cost more in time and labor than the settlers really could afford. The frame building also was poorly adapted to housing livestock,

Barn builders were attuned to their environment. They considered the contours of the land in designing their structures.

RIGHT: The growth of this Vermont farm and the changing range of its operations can be tracked through the additions on both sides of the barn.

OPPOSITE: Barns today are often a combination of different styles developed and adapted over time to meet particular needs.

A very early American log barn, circa 1790, has been reconstructed as part of the Landis Valley Farm Museum in Lancaster County, Pennsylvania.

which was much more critical in the harsher American environment than it had been in England. And there you have part of the reason that early English settlements in the New World attained only checkered records of success.

By contrast, the first French pioneers of southeastern Canada were quick to adapt to their new surroundings, both in their buildings and their attitude toward this new land. To build their barns, they laid logs one atop another and secured them by corner posts. This quick method required the least amount of labor.

Most of America's first barns were of log construction, with mud or straw sealing the openings between logs, and thatch, sod, bark, or more wood forming the roof.

The Swedes and Finns, who arrived a bit later than the British and the French in the region around modern-day Philadelphia, were well prepared for construction in North America. They had heard the tales of woe from previous would-be settlers who had returned in failure to Europe. But, more important, their traditional style of frame building with logs was well suited to this new land. The needed materials were abundant and the construction techniques adapted easily. Their style of building also allowed for erecting larger barns than could have been made from simple log construction.

Their practical natures gave precedence to the construction of fine barns, well before their own homes came anywhere near the same level of finish.

Two other groups of pioneers also came to the New World equipped with basic barn designs that could be easily adapted to this new environment. They were the Germans, who arrived in southeastern Pennsylvania in the 1680s, and the Dutch, who came to occupy the river valleys of New York around the same time. With the rich soil they found in their new environments, the farms of these two groups flourished and became models for settlers moving into the wilderness to the west, north, and south.

The log barn at Abe Lincoln's farm in Coles County, Illinois, provides a prime example of the construction methods employed in the region at the time the future sixteenth president lived there.

OPPOSITE: Logs laid on top of each other and secured by corner posts to form a snug enclosure was one of the earliest and most efficient methods of barn construction in colonial America.

Although the traditional designs, such as the Dutch barn, have persisted through the decades because of their practicality, many alterations have been attempted over the years.

NEW WORLD INGENUITY

The Dutch were particularly adept at adapting their traditional barn style to the North American environment. Their barns—large, two-story, wooden structures with spacious haylofts above a working floor, and with aisle arrangements of stalls for the livestock—became the foundation of the American style of barn.

That style was definitely modeled after European precedents, but with a big difference for many of the cultures that would adopt it in this New World. The barns of much of Europe were small affairs, only as roomy as they needed to be to house the limited livestock of a typical farm. But the New World was the land of opportunity. The farms the settlers carved on this land most certainly would grow beyond anything they'd seen in their native Europe. And such growing farms would need plenty of barn space. With that thought, the settlers built larger barns than the world had generally seen before. Such barns were among the very first unique creations in American architecture.

Working from a purely functional viewpoint, the pioneer barn builder began with that most basic of shapes—the square, or several squares connected side to side. Roof design was equally simple: Draw a straight-edge from the corner of one square to the corner of the appropriate other square and you had an angle that would shed rain, snow, and sleet quite effectively. (As with any "rule," however, we will see later that this one too has been broken.)

In that simplicity lies the motivation behind the universality of design we have in our barns today from coast to coast. The concept of squares, corners, and lines was easily remembered, easily copied, and readily put into action. As European settlement spread westward across the continent, that concept was carried with it. It even found its way into the emerging styles of homes and other buildings that the Americans eventually built.

The huge doors to the second level of this Pennsylvania-style barn provide ready access for the equipment that generally is stored on that level.

EARLY OUTBUILDINGS

One of the earliest outbuildings to appear on the American farm, after the barn, was the forge barn, a small building with a chimney that was built near the larger structure. Here the farmer, working with a stone forge, could create all the hardware and tools he needed while avoiding the British taxes on them. The bellows for these forges often were made from the hides of deer, elk, or moose. The forge barn lost its usefulness as the colonies became more settled (which meant that a larger community could support a specialist blacksmith) and with the removal of the taxes on hardware and tools. At this point the farmer generally converted the forge barn into a toolshed.

Another outbuilding, the corncrib, actually saw its origin as a corn barn. And, as American as the structure may seem, it actually originated in an English plan. To foster air-drying of the corn, the corn barn was erected on a series of posts, which were further adapted with outward projections to keep rats out of the barn. The roof was equipped with a series of wind vents.

Although the corn barn was short-lived in America, where it never gained wide acceptance, its smaller cousin, the corncrib, survives today. The earliest corncribs—simple, square log lean-tos with stockaded sides—were recorded as early as 1700. A series of adaptations developed the crib into the slatted, outward-slanting structure we know today. The vermin-proof post idea of the corn barn was carried over.

Many corncribs included two bins with a drive-through area between them, all under a common gable roof. These often were the first structures on the mountain farms of Appalachia, later evolving into the cantilevered barns of that region.

By necessity, farm buildings have always adapted to their surroundings. Despite its name, the Waterville area of eastern Washington state is notoriously hot and dry. To survive, a local farmer located a water tower and pump house next to his barn.

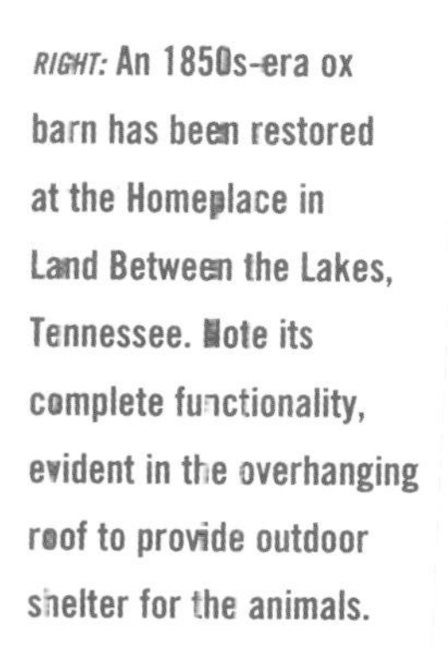

RIGHT: An 1850s-era ox barn has been restored at the Homeplace in Land Between the Lakes, Tennessee. Note its complete functionality, evident in the overhanging roof to provide outdoor shelter for the animals.

OPPOSITE: This corncrib in the Great Smoky Mountains features the classical log construction typical of farms in the region.

LEFT: The exterior hayloft access is a relatively recent development in American barns. Earlier barns provided this access from the central work floor inside.

OPPOSITE: The end of the peak of this barn features a variation on the top-hat design, which originated in tobacco barns. Here the design feature is employed to provide shelter for the loft door.

HAYLOFTS AND SILOS

Haylofts, one aspect of barns that nearly everyone can identify, were not a part of the first barns in America. On those tiny, pioneer farms the work was done by hand and any meat for the table was bagged with gun or trap. Livestock was simply not kept. Those first barns were used to store what little grain the farm could produce and as work areas, primarily in threshing that grain.

It was not until the farms began to grow and prosper that more specialized buildings were needed. Barns with special purposes then began to appear to house everything from chickens to oxen, to store various types of farm production, and to perform the normal tasks that make a farm function.

Silos, which today we have come to expect to see next to the barn, were one of the relative latecomers in the succession of specialized farm structures. It was well through the 1800s before these particular structures began to appear.

BELOW: **The artistry and perfect proportions of this wooden silo in Washington state bespeak the many long hours of labor that went into its completion.**

OPPOSITE: **It wasn't until the 1800s that many of today's now-standard basic farm structures, such as the silo, began to appear next to the barn.**

FOLLOWING PAGE: **Pigeons, like these perched on a traditional New England barn, have been a fixture on barn roofs since the first farms were cleared in America. On the very first American farms they were the principal source of fowl.**

CHAPTER TWO

FORM AND FUNCTION

The German settlers in Pennsylvania—commonly known as the Pennsylvania Dutch—took the barn style of the Dutch yet another step closer to its classic American configuration. The first significant change was the addition of a third story. It was built into the bank of a hill, hence the names bank barn and sidehill barn. On the side of the barn opposite the hill the second level traditionally overhung the lower level, providing a sheltered area over the entrances to the animal stalls. However, a common variation instead had pentroofs extending out in place of the overhanging second level. In these barns, the floor joists of the second level were allowed to jut out away from the building and the pentroof was built on them.

The overall size of the Pennsylvania barn continued to increase as farmers became better able to grow and store the hay needed by their increasing livestock inventories.

The tight mortar seals between the stones of the foundation of this barn ensure that the lower level naturally maintains a relatively constant temperature throughout the year.

RIGHT: This Bucks County barn displays all the traditional features of the classic Pennsylvania structure. The principal difference between it and its predecessors is that it isn't built into the side of a hill.

OPPOSITE: The simple weather vane, not the ornaments that currently fill antiques shops, was a critical instrument to the early American farmer, whose work was entirely dictated by the weather.

An amalgamation of barn styles and features—gable roof, vented cupola, and stone foundation—have been brought together in this barn near Polk, Pennsylvania, to serve a variety of needs.

BUILDING FOR WINTER

Building the farm into the side of a hill brought many benefits in fighting the vagaries of nature. Barns have always been situated to take best advantage of favorable weather and thwart the impacts of inclement conditions. Prevailing winds in the region, the tracks of the sun through the sky in the different seasons, the lay of the land on the site, weather-impacting landscape features, and myriad other factors have always been considered, studied, and plotted before any ground is broken. Wind direction was generally paramount among these considerations.

Design features that mitigate the effects of cold winter winds are obvious in even the earliest barns. The saltbox barn, which was one of the earliest American designs, features a roof with one much longer side that slants from the peak almost to ground level. This longer side of the roof faced the prevailing winds of winter, which generally blow from the north. For this reason, the longer side came to be known as the "north roof."

The space from the lower end of the north roof to the ground was packed with hay, cornstalks, leaves, and the like during the fall. When winter arrived and the snows began to fall, they would pile up from the ground right onto the roof, providing insulating warmth.

The weather is critical to everything that happens on the family farm, and winds are critical to much else that happens with the weather. So the weather vane atop the barn, where it could readily catch and report on those winds, became a fundamental instrument on the American farm.

Despite their seeming ubiquity in antiques shops, highly ornamental, figurinelike weather vanes are actually a relatively recent addition to the American barn. The first several generations of farm families in America were relatively religious and, for many of them, decoration was taboo. As a result, their weather vanes were more likely to be simple, flat, wooden pointers or pieces of cloth that flapped in the wind than prancing horses or crowing roosters.

It was not until the mid-nineteenth century that the more ornate weather vanes really came into vogue on barn roofs. Only then did the various shapes that we now view as traditional become commonplace. Concurrently, those weather vanes that incorporated motion, such as a bird with flapping wings or a pair of lumberjacks sawing a log, began to surface. The most elaborate weather vanes actually sacrificed some of their wind-reading viability for decoration.

LEFT: The cantilevered design of this traditional Pennsylvania barn provides a sheltered area for the lower-level livestock area with the second-level overhang.

OPPOSITE: The saltbox barn was designed to provide protection from the cold winter winds. Usually facing north, the long end of the roof was packed with hay and cornstalks to provide insulation for the barn's interior.

A connected style of building became popular among farmers in the more northerly climes, allowing them to pass from one building to the next without exposure to the weather.

CONNECTED BARNS

In addition to weather-determined positioning, the barn has generally always been situated in a central location on the farm to minimize the amount of travel between it and the other buildings. In some of the more northerly climes this penchant has developed further into what is known as continuous architecture. Here, when the snow falls and the temperatures plummet, the farm family can move from the home to the outbuildings to the barn without ever emerging from the cover of a roof. In a typical scenario, the barn adjoins the utility shed, which is connected to the milking room, alongside the summer kitchen, just off the living quarters of the home. The exact order of connection would vary with the preferences of the farmer.

The white barns of Canada, also known as long barns and connected barns, bear a similarity to the connected barns of New England. However, they were modeled after those of France's Normandy region by specific order of Samuel de Champlain. When Quebec became so overpopulated that pasture had to be developed at some distance from the city, the governor ordered that a satellite village be established at Cap Tourmente. And the first requirements in erecting that village were the barns for the animals, which Champlain specified should be "60 feet long and 20 feet wide . . . made of wood and earth, like those in the villages of Normandy." Those farms of Normandy consisted of several outbuildings and the farmhouse all separated around the complex,

As the central structure in American landscapes, like this in the Pennsylvania Dutch country, the barn creates an aura of strength and permanence.

LEFT: In keeping with their lifestyles, the Amish build and maintain their barns simply and with little ornamentation. The clothesline, running from farmhouse to barn, is a regular fixture on their farms.

BELOW: The Pennsylvania Dutch country in the south-central region of the state is a photographer's wonderland of picturesque scenes, but these well-kept barns and farms are most definitely working entities.

and the first of the Canadian long barns followed that plan. However, the quick-to-adapt French Canadians soon found that a connected plan, similar to that in New England, had distinct benefits in the harsh winters of the Quebec province.

BARN PROTOTYPES

The Pennsylvania standard barn and the Yankee barn of New England provided the prototypes for many of the other barn styles that developed as pioneers continued to push the realm of settled North America toward the west. Using stone and red brick for the side ends of the barn, the Pennsylvanians built their barns into the sides of hills with their livestock stalls on the lower level that receded into the hillside. Several Dutch doors opened from the lower level into the barnyard. The second level, which could be entered directly from the hillside, was the machinery floor. With this layout a great deal of hoisting to the second level was eliminated. On the side of the barn opposite the hill, this second level overhung the lower level with what was known as the "overshoot." The

The original Pennsylvania barns were built into the sides of hills to fully utilize the ground's natural insulation factors, as well as to provide easy access to upper levels.

OPPOSITE: The Yankee barn is usually constructed of wood instead of stone. While early barns of this type had gable roofs, later builders favored the gambrel roof to provide additional space.

Dutch doors, common in the lower animal-housing level of barns, provide ventilation while still confining the livestock, as demonstrated by this billy goat.

loft topped off the structure, which was completed with a gable roof.

Mennonites followed the basic Pennsylvania standard, but with a passion for permanence. Only oak was strong enough wood for their use and, wherever they could, they used stone when others used wood. The walls of their barns often were built to incredible thickness. Is it any wonder that these barns are still in remarkably good condition?

The Yankee barn was quite similar to the Pennsylvania standard, but the New England environment provided more wood than workable stone or clay for bricks and so the stone or brick sides were replaced with wooden boards. In another bow to the environment, the New Englanders employed the English gambrel-type roof to better shed the heavy snows they encountered.

One apparent variation on the English barn of New England is the double crib barn of the Appalachian Mountains, from southern Pennsylvania southward. With two similar cribs separated by an open access and work area, all topped by the loft area cantilevered out beyond the base on all sides, the double crib barn allowed the farmer to carry out nearly all of his winter chores without ever leaving the sheltered area. One of the cribs generally held cattle, while the other housed horses.

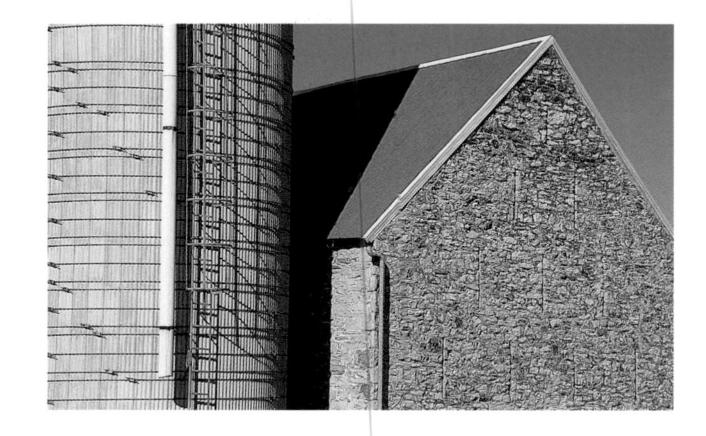

The Mennonites of south-central Pennsylvania built their barns to last. Whenever possible they built with stone rather than wood.

OPPOSITE: Though not set against a hillside, this barn in Lycoming County, Pennsylvania, is typical of the Pennsylvania style, with stone side ends and wooden upper floors.

The feed barns of the southern Appalachian Mountains feature storage bins on each side of a central access and work area that runs through the entire structure.

The slatted side boards of the typical Lancaster County, Pennsylvania, tobacco barn pivot out to create openings in the sides of the barn. These allow air to pass over the hanging crop inside.

Tobacco barn racks are designed to hold thousands of hanging Nicotiana leaves in the minimum amount of space required for adequate air passage over the drying crop.

TOBACCO BARNS

The Amish of Lancaster County, Pennsylvania, and farmers to the south raised a crop in the New World that required an entirely different type of barn structure. Tobacco was a very valuable crop, but it required special handling to bring out its worth. The harvested leaves couldn't be simply stacked like bales of hay or straw, nor could they be poured into some watertight vessel as with grains. Instead, they needed to be slowly and naturally air-dried if they were to be worth anything.

The tobacco farmer responded to the special needs of his crop with the tobacco barn: A long wooden structure with a gable roof, which opened at its peak under a smaller gable roof to allow for additional ventilation. For added ventilation, the sides of the barn were raised a foot or two off the ground on regularly spaced "legs" and there was no floor. Along the sides of the barn, every other board or every other pair of boards was hinged to allow for vent openings the entire length of the barn.

Interiors of tobacco barns have a skeletal look. Beams extend in from the walls the whole length of the barn, leaving a work aisle along the center. Onto these beams racks of inverted tobacco leaves are hung to aircure.

The top-hat barn is a variation on the tobacco barn. The gable roof has been modified with large rain-hood

extensions over large wide openings at both ends. In addition, the swinging side boards are replaced by full windowlike openings.

Just as the farmers of the tobacco-growing regions came up with specialized structures geared to specific crops, so did the Yankees of New England. Their special crop was sugar from the maple tree, and their specialized structures came to be known as sugar houses.

At first the maple sap was simply boiled down in large iron kettles out in the open without benefit of shelter. However, in the early 1800s, when tin evaporation pans came onto the scene, something was needed to protect those relatively expensive tools from the elements. At first it was just a simple cabin, designed to allow the workers to be present around the clock for the month during which the boiling took place. Later it evolved into the specialized sugar house with roofed ventilation openings and covered areas for the extensive supply of firewood needed to maintain the fire.

BELOW: **Barns on the farmsteads of the Blue Ridge Mountains tended to be rather small until quite recently, attesting to the effort and energy needed to eke out a living from the land in this environment.**

OPPOSITE: **A typical tobacco barn in the mountains of North Carolina displays the open-air drying compartments and hanging tobacco-leaf racks so critical to successfully curing the crop.**

FOLLOWING PAGE: **Air-drying on the racks of this traditional Southern hill-country tobacco barn, a leafy crop cannot be hurried to market.**

TOPS
TOPS

Individual stall entrances, both from the outside and the interior of this horse barn, provide for great efficiency in working with the livestock housed there.

A traditional horse barn houses thoroughbred livestock on this farm near Palo Alto, California. The bank of windows along the roof extension provides for a well-lighted interior.

HORSE BARNS

Another development in the evolution of barn design and construction that arose as farms moved from general, multi-crop, agricultural production into more specialized methods was the horse barn. Reflecting the extremely high value of its livery occupants and the careful attention they receive, the horse barn is big, roomy, airy, and easily cleaned and maintained.

Individual stalls, each with its own doorway to both the outdoors and the interior area, line both sides of the barn, flanking a central work area. The interior side of each stall, and often other sides as well, is solid wall only to a height approximately even with the back of a horse, topped with vertical bars, allowing for an occupant-calming view beyond the stall and healthy air flow through the confined area. Feed and water troughs are built in, often to the interior side of the interior wall and usually in such a fashion to allow the horse access to the food and water but preventing spills.

A separate tack room replaces one or a couple of the stalls in a corner or at a central point, providing both an area for storage of tools, equipment, and harness and something of a headquarters for the operation.

Nearly all the accoutrements of thoroughbred horse farms, including the barns, reveal how this type of farming differs from more traditional farms.

CHEW
MAIL
POUCH
TOBACCO
TREAT YOURSELF
TO THE BEST

CHAPTER THREE

CUSTOMS AND TRADITIONS

Early in the nineteenth century, as America's ability to plane its raw timber into boards developed, frame construction was coming to dominate the American barn-building scene. Gambrel roofs were now popular, both for the ornamentation they added and for the additional loft space they afforded. Similarly, masonry was becoming nearly as common as the stone foundation and side works that had been the standard for so many years.

PAINTED BARNS

Painted barns began to appear on the American landscape in the late 1700s. Prior to that time farmers practiced weather-curing on their wooden structures. Being intimately tied to the land, the wood, and the weather, they learned the precise connections necessary to erect a barn (or house or other building) of green timbers so that as those timbers cured they would hold fast and become even more secure on the structure. We can see how far we've moved away from the natural world in which these men and women lived if we consider that the appearance given to a barn by weather-curing is often referred to today as "weather-beaten." So, until the late eighteenth century, paint on a barn, house, or other building in pioneer America would have been viewed as a useless and frivolous waste of scarce time and energy.

But by the late 1700s the farmers in the more settled areas of the continent had carved away enough of the wilderness to take time for some extras. Weather-curing, while a true art, was not a science. With unfortunate regularity, it would leave problems to be repaired on the barn. Paint was the replacement of choice for the old method.

Virginia farmers were the first to paint, developing a coating that drew its gray tones from lampblack. Their more northerly counterparts moved in the direction of red, creating a mixture of the rust from iron, skim milk, and lime that would harden into a near-plastic coating

RIGHT: The traditional "barn red" coloring that we expect to see on American farm buildings was originally a mixture of the rust from iron, skim milk, and lime.

OPPOSITE: The familiar Mail Pouch chewing tobacco signs, painted directly on the sides of barns across the country, are becoming even rarer than the barns that sport them.

on their buildings. Eventually they added linseed oil to give the paint the ability to soak into the wood on which it was applied. Thus was born the "barn red" coloring that came to be the standard for barns across the country.

The red barn is a familiar icon of the rural American landscape, but the average traveler's eyes would not be shocked to also spot an advertising message spread across the side of a barn. Although the practice of painting advertisements on the sides of barns hasn't been common across much of America for some years now, the image is still very much a part of our collective consciousness.

One of the greatest practitioners of the barnside advertisement was the Mail Pouch chewing tobacco company. Even those of us who have never stuck a wad

ABOVE: **Medicine purveyors were quick to recognize the advertising potential that the large sides of barns offered.**

OPPOSITE: **Though today some might call the boards on the side of this barn "weather-beaten," traditional barn builders regarded it as perfectly weather-cured.**

Although many barns today feature intricately detailed artwork, even some by top artists, in general these utilitarian structures did not carry any paint at all until the late 1700s.

of tobacco in our cheeks know the name of the company because of this practice. Even after the company stopped paying for the paint on some barns, local artists sometimes took up the practice to maintain what had become a landmark of the home territory.

The huge, broad sides of barns, which generally were found alongside or at least well within view of the highway, held an irresistible attraction for many early advertisers. Patent medicine companies were heavy users of the medium.

In exchange for allowing use of the space, the only payment the farmer often received was a fresh coat of paint for the barn and a sample of the product being featured.

The bright red barn on the Daniel Boone Homestead in Birdsboro, Pennsylvania, did not exist while the legendary woodsman lived on the homestead in the 1730s through the 1740s. And, if it had, it most certainly would not have been painted red.

Painted murals are also sometimes seen on barn sidings. Though elaborate scenes on barns were common in Bavaria and Austria, American pictorial efforts have traditionally focused on livestock, particularly horses. In recent years, barn artists have painted vivid designs that usually express a patriotic theme.

Another instantly recognized feature of barns is the hex sign of the Pennsylvania Dutch, which traditionally featured some geometrical star pattern within a circle or concentric circles on the side of the barn. One theory, by far the more popular one, has it that the Amish farmers painted the hex signs on their barns to ward off the evil influences of witches or Satan. And, in point of fact, the German word for witch is "hex." Another theory is that hex signs are simply decorative folk art that the farmers carried over as a remembrance from their German homelands. Scholars have debated this for decades, and continue to do so.

BELOW: With its grand mural and silo swirled like a peppermint stick, this Minnesota building certainly is in the running for the most colorfully painted barn in America. It also features sophisticated cupolas atop the roof for ventilation.

This three-dimensional wooden construction is based on traditional hex signs. The real things were more traditionally painted onto the sides of barns in bright, primary colors.

Besides the Amish, others also take advantage of the group method to quickly erect a barn. Once assembled, the wall units are then raised, as all the workers join efforts.

Working quickly, the group affixes the wall units into place and connects them to form the skeleton of the barn.

AMISH BARN RAISINGS

Another time-honored tradition among the Amish, and in past times some other rural American groups, is the barn raising. Only a few groups, notably the Amish, continue the practice today. And when one of the Amish communities, such as those spread throughout Pennsylvania and Ohio, get together to raise a barn it's a sight to behold. Dozens of darkly clad men swarm over the site, first building the walls on the ground, and then, when enough able bodies have arrived, raising and affixing them into place. By the time much of the outside, modern world is just starting the day, a great deal of the project has already been completed. Meanwhile, the women gather around the farmhouse, setting up long, white-clothed tables on the lawn and covering them with some of the best home cooking you're ever likely to encounter. When the work is done the builders celebrate with a communal feast.

ABOVE: The barn-raising today continues mostly as a tradition among the Amish. The raising begins with the cutting and notching of the wall components and the assembly of the full walls, all done at ground level.

With the barn's frame in place, a swarm of workers fill in the structure with the roof and side boards.

Construction of round barns became much more practical after American builders had fully developed frame construction techniques.

Utopian societies, such as the Shakers in New England, originated the round barn design in an attempt to create more heaven-connected structures here on earth. Indeed, this barn has the ecclesiastical feel of a church.

SHAKER IDEOLOGY

Round barns, and their succeeding octagonal barns, arose largely out of the desire by utopian societies, such as the Shakers, to create more heaven-connected structures here on earth, but also because the emerging abilities in frame construction in America made them possible.

To the credit of their devotion to heavenly pursuit, the Shakers were able to erect a ninety-foot-diameter round barn near Hancock, Massachusetts, as early as 1826. The conical roof, complete with separately roofed window banks on all sides, was topped by a louvered cupola. A central, hollow ventilation opening ran from the base of the barn to the cupola.

Although some modern-day gentleman farmers have been lured to the round or octagonal designs, and a few day-to-day farmers have chosen them for their efficiencies, this design was never common on American farms. It remains a traffic-stopping rarity today.

Proponents of the round barn tout the circular design's ability to encompass the maximum amount of floor space within the minimum amount of walls.

FOLLOWING PAGE:
Simple yet functional has always been the primary requirement for the workings of the American farm. The straightforward mechanism of the barn-door bar continues to be employed today.

CUPOLAS

Resting atop many barns we find additional small structures, tiny squares with slat-covered sides often topped with weather vanes and/or lightning rods. These are known as cupolas, a misnomer that applied to only the very earliest of their kind. Named for the cups they resembled, the first turretlike, dome-shaped cupolas were intended for viewing the outside world.

However, the structures were soon put to a much more practical function, providing much needed ventilation for the barns on which they perch. The hot, stale air that collects in barns even during the coldest periods of the year pushes out through the vents of the cupola, while the elements outside are still held at bay. The use of cupolas to this end originated in the New England states, soon after farmers began the switch from a subsistence form of agriculture to a more market-oriented approach to their operation.

Lightning rods are a critical feature on barns, which often are the most outstanding and tallest feature in their rural settings.

OPPOSITE: Cupolas, some simple, some ornate, were added to barns as ventilation devices to remove some of the heat and stale air that can collect in the upper reaches of such large structures.

While most of the other features of barns are very similar from one barn to another in the same type of barn, the cupolas became an opportunity for self-expression for many farmers. Regional influences were strong for some, such as the builders of the many lighthouse-like cupolas throughout the coastal areas of New England. Additional uses were part of the design decisions for others, the reason so many cupolas include extra areas with openings for pigeons.

But for others, notably the "plain" societies like the Shakers and the Amish, the cupolas were just another functional part of the barn. They were kept simple and without any extra ornamentation, with just enough design to fulfill their purpose.

Gothic in style, this green shuttered vent is a beautiful yet utilitarian feature on this Pennsylvania barn.

Cookie-cutter detailing came to the American barn, as it did to much of American architecture, during the design-conscious Victorian period.

OPPOSITE: Wooden cutouts in the sides of barns helped prevent heat from building up within. The decorative ventilation work on this circa 1890 to 1900 Pennsylvania structure attests to the Victorian love of ornament.

Slatted vents, such as in the cupolas atop this barn, provide ventilation for the structure while at the same time sheltering it against the elements.

THE VICTORIAN AESTHETIC

As America moved into the second half of the 1800s, the Victorian penchant for building ornamentation began to creep into the architecture of barns as well. Not that everything that Victorian America brought to the advancement of the barn was frivolous. No, all the advances in building technology—from materials to ventilation to plumbing—were also being incorporated into the barns. But much more obvious was the facade, complete with tower and spirelike cupolas, cookie-cutter trim, and larger expanses of windowed walls.

Every feature of the barn—roof lines, cupolas, doorways, windows—presented the designers and builders of the period with additional opportunities to "show off" the prosperity of the age, and of the barn's owner. As in all areas of American architecture at this point in our history, barns became the focus of contests, both formal with prize offerings and informal among neighboring farms.

Large barns were elevated to almost cathedral status in their finery. Smaller barns took on a decided cottage appearance. Additional windows, never envisioned for any reason in earlier ages, now became ornamental necessities. In some instances the traditional functionality of the barn was forced into a back-seat position to the new aesthetic considerations.

Completely new barn styles were also being experimented with during the period. The multi-winged, compound-like structure, radiating out from a central work area, was one of the popular new designs. The old concept of connected barns from New England and Canada in an earlier day was revived with new respect and a new attitude.

On the more practical side of advances during the Victorian Age, barn builders could incorporate such "newfangled" contrivances as metal housing for doors that allowed them to slide smoothly open and closed or windows with movable panes.

The Victorian age coincided with a period of prosperity for many American farmers. The more ornamental architecture of the period allowed those who were so inclined to make new statements about that prosperity.

LEFT: Although it is still quite picturesque, this weather-bleached Massachusetts barn shows the telltale signs of age that have claimed many old structures.

OPPOSITE: The ornamentation around the cupola of this New York barn is typical of the flair the Victorian period brought to the farm.

1881

CHAPTER FOUR

EVOLUTION

The barn has been a source of pride for the farmer from the very earliest constructions. While the creation dates for most farmhouses were generally not recorded in any formal manner—at least not until the advent of modern building permit requirements and the like—the date that the barn went up was a date of monument. Not only did the farmer note it precisely in his business ledgers, he also painted it clearly for all to see at the peak of the barn roof. The barn was the center of the family's farming business.

The additions to the sides of this barn are known as jogs. They were built as the needs of the farm grew and, thus, reveal a bit of the history of the farm.

Glass was a luxury beyond the reach of most early American farmers, even for the first homes they built. It was virtually unheard of in barns prior to the eighteenth century. Opened doors and permanently open, slanted louvers were the only means by which sunlight and fresh air were allowed into the barn. Occasionally glass bottles were fitted into an opening in the side of the barn, held in place with clay, to provide additional lighting.

Ventilation also was provided by pigeonholes cut into the wooden walls of the barn. These holes also served to attract the flocks of wild pigeons that the pioneers encountered in great profusion across the early American landscape. Although today we think of the

RIGHT: Whether on the prairie or close to the sea, the relentless wind has always determined barn design. A long, sloping roof, facing into the prevailing winds, is the traditional solution.

OPPOSITE: Gambrel roofs, such as on this barn near Woodstock, Vermont, originated as an adaptation of early American farmers to the heavy snows they encountered in New England and the Mid-Atlantic states.

The barn swallow house on the side of this barn is reminiscent of the pigeonholes that early American farmers cut into the sides of their barns to attract wild birds for the table.

chicken as the stereotypical farmyard bird, the early farmers preferred the gradually semidomesticated pigeons. They were a staple of the early American farm diet, replaced—on special occasions—by a wild turkey or wild duck.

GABLES AND GAMBRELS

One of the most distinctive features of any older barn is its roof. The design of that peak speaks volumes about the models that the builder chose to follow and about the life of the barn. The simple gable—two equal sides to the roof meeting at the peak—was the earliest design attempted in America. However, it was soon replaced with the saltbox, which borrowed ideas from the native Americans about using the insulating abilities of snow to add warmth and comfort to the barn. The saltbox features a longer side, often known as the north roof, which is faced against the prevailing winds of winter.

The snug Dutch roof style is a variation on the gable, which adds slated snub-nosed roof areas at the ends of the roof, which would end flat in the gable. Similarly, the English gambrel and Dutch gambrel roofs are advanced variations on the gable. They both add an extra bent to the roof, creating a gradually sloping roof area extending down from the peak followed by a much more abruptly sloping area reaching the ends of the roof. The Dutch version adds a slight convex curvature to the steeper slope portion of the roof. These variations on the basic gable obviously would speed the dropping of accumulated snow from the roof.

Although in later constructions they were incorporated into the original design of the barn, several roof features in older barns reveal additions to that original design. Principal among these are the hip, jog, end gable, and broken gable. Although today any of these might be incorporated into the original plans for the barn out of ornamental or architectural concerns, in the past there were much more functional reasons for them. The hip, generally seen as an extra, outward-slanting roof at the end of a gable roof, in times past signified that the portion of the building it covered was an addition to the original building. Similarly, the end gable—a miniature gable roof atop a smaller, shorter addition to the original barn—signaled growth of the facility. The same applies to the jog, which is a lean-to roof over a smaller addition to the barn.

Two classical barn roof designs are displayed on this Michigan farm. The barn at left is covered by the Dutch gambrel roof, while the barn at right sports an English gambrel roof.

Almost a signature piece for the Midwest, a modified Dutch gambrel-roof barn is surrounded by the farm's fields of grain.

This barn in the Midwest has incorporated the sloping-roof principle of the original New England saltbox barns to help shelter the barn against the fierce winds of the region.

OPPOSITE: This jog addition to a Vermont barn demonstrates the sloping-roof design, albeit with a slight broken variation, that was typical of the early saltbox barns of the region.

A traditionally styled Western ranch barn near Helena, Montana, displays a design from the late 1800s, including metal roof and cupolas.

WAY OUT WEST

As the first settlers moved onto the plains they ran head-on into a completely new and different environment. Here there was very little wood, the flat land offered little environmental protection against the elements, water sources were not dependable, and their products—mostly cattle at first—were different. Pretty much of what they had learned in the East was useless in this new land.

Winter shelters for their cattle, built of poles and straw and hay, were the first "barns" of the West. Throughout the winter, after each storm, the farmer/rancher would heap new straw and hay onto the structure to replace what the animals inside had eaten and to increase the insulation factor.

Variations on these shelter barns included the following. The Kansas tepee was an erected A-frame of poles at the center with corral-like fences

RIGHT: As America pushed its borders to the west, many new barn styles and adaptations were conceived in the pioneering spirit. The round barns that originated a century earlier with the Shakers in New England were given a new life, as with this structure near Great Falls, Montana.

OPPOSITE: In the majestic, mountainous landscape of the West, such as Antelope Flats at the base of the Grand Tetons in Wyoming, the barn sometimes takes on a much less imposing appearance than it would have in less grand settings.

FOLLOWING PAGE: The traditional prairie barn features stalls for livestock on either side of a central work and storage area.

on either side. Cattle were housed under the A-frame, fed and insulated from the hay and straw heaped into the corral-like areas and atop the A-frame. The lean-to shelter was what its name implied, a lean-to pole frame with a heap of straw and hay stacked atop it for the animals housed under it. The Western straw shelter involved a long, U-shaped arrangement of connected arbor poles, over which the straw and hay was heaped.

Field barns in the West were an adaptation of the silos in the East, responses to the incredible dryness that persists through much of the year on the plains. The field barn began as a circular corral-like structure of stanchions, with height added by an additional circle of wire fencing material. The hay was heaped into the enclosure for the cattle to feed on through the stanchions. A roof of straw or shingles sat atop the wire fencing material or was propped in a slightly raised position over the structure with a system of boards.

The huge expanses of roof on the traditional Western barn, such as this example in Utah, were the rancher's response to the harsh winds and heavy snows of the region.

RIGHT: Various visually intriguing roof designs have been developed over the centuries to provide ventilation for the barn while still sheltering the interior from rain, sleet, and snow.

OPPOSITE: The use of a second, smaller gable-roofed structure atop the barn roof originated with tobacco farmers, who were looking for yet more methods to ventilate their crops.

The round barn design requires a very different system of beams and rafters than that of the more traditional and widespread rectangular layouts.

Still standing after all these years, an old homestead near Telluride, Colorado, is a poignant reminder of the challenges of Western pioneer life in the late 1800s.

The garden behind this restored barn shows the spectacular results that can be achieved by turning these old structures into homes or businesses.

All of these were rather temporary shelters, both in appearance and application. They could be disassembled and moved to other locations quickly. This was another of their adaptations to the new environment. As the settlers' ability to tame their surroundings advanced, the way was cleared for buildings of more permanence. Experimentation with new styles also seemed to be in order in this new land.

When the westward-ho pioneers did get around to erecting more permanent barns, they found that the vicious winds that swept across the prairie were their greatest enemy. Their barns reflected this with a number of wind-deflecting designs. These included such widespread ideas as a modified version of the top-hat tobacco barn, in which the top hat was intended to cut the winds, and a version of the saltbox barn, with its "north roof" facing the direction of the prevailing winds. Their barns took on much greater dimensions than the models on which they were based, but the same principles prevailed.

Round and octagonal barns were given yet another chance to prove themselves in this atmosphere. Unfortunately, they often did not fulfill their promise. The basic theory behind the round barn was that a circular design would encompass the maximum amount of floor space within a minimal amount of walls. While that was true, it led to problems in maneuvering hay and straw into position for maximum storage and, of course, the pie-slice-shaped livestock stalls created a lot of unused space.

BARNS TODAY

The barn, whether round or square, remains central to the farm, but with much less sentimentality than it once carried. Modern barns, as with other aspects of modern architecture, are statements made with steel and concrete. Only the sentimentalist or recreationist today builds a barn of the traditional American style and completely of wood.

The modern addition of concrete not only made many aspects of working in and cleaning up around the barn easier and quicker on the farm, but also brought sanitation to a higher level. It was this latter quality that led many regulatory agencies to institute requirements for the replacement of stone foundations and ground or gravel floors with concrete. Similarly, galvanized metals became the material of choice for the roofs and sides of barns because of their ease of installation and their long-lasting qualities.

The most modern of trends pertaining to the

Though this Vermont barn is being constructed along traditional lines, its roof and siding will be completely modern to insure durability and easy maintenance.

Barns are popular conversion subjects these days. This round barn near Lexington, Kentucky, has been refurbished to create a rather unusual restaurant setting.

American barn is the conversion of the old structures into homes, offices, and retail space. The large size of the existing buildings lends itself well to the creation of incredible living spaces and multiunit working and shopping areas. The huge beams that frame the barn remain in remarkably serviceable condition. And the same construction designed to keep the animals comfortable and the stored grains or silage in mint condition performs equally well to keep human occupants comfortable.

While many barns have not met with such renewed uses and barns in general have been disappearing from our countryside at an alarming rate, the picture for the future is not entirely bleak. Several organizations have come forward to address this problem.

Few organizations have done more in the area of rural conservation over the past decade or so than the National Trust for Historic Preservation. It even has a regional office in Denver that has focused on helping farmers and ranchers to find new ways to use their aging barns.

In a joint effort with *Successful Farming* magazine, the Denver office has created a program called Barn Again! Through a contest, which drew entries from more than five hundred farmers in thirty-four states, barn rehabilitation was encouraged nationwide. The program offered advice on refurbishing barns and presented prizes of up to one thousand dollars each for prime examples.

Most refurbished barns on working farms are brought back into service for normal farm chores, but some are redesigned into completely new, nonfarm uses. Antiques shops and other stores are a few examples. The large, sheltered expanses of barns also lend themselves well to theatrical and group-activity facilities. The back-to-the-earth feeling that comes with housing one's business in a barn has attracted artists from many fields.

Tourism, too, can be a strong motivation in preserving and restoring old barns, particularly for governmental and quasi-governmental agencies. Many states have included old farms, complete with their barns, in their historical site programs, inviting the public to come and view a way of life that's fast vanishing.

Although we've lost quite a bit of our barn heritage that can never be replaced, thanks to efforts such as these, that icon of Americana will remain a part of the American landscape.

LEFT: Wonderfully weathered wood, such as the planks on this barn door, has become a coveted component in many home redecoration projects, particularly when a country feel is the goal.

OPPOSITE: A Western hay barn keeps the worst of the rain and snow off the hay while it provides ample air circulation with its open sides.

PHOTO CREDITS

***Photographer**/Page Number*

American Landscapes
Ray Atkeson 23, 76 (top)
Rick Schafer 79

Charles Braswell, Jr 4, 15, 18, 36 (top), 38, 39

Bullaty Lomeo 21, 27 (top & bottom), 30 (bottom), 31 (bottom), 40-41, 47, 62 (bottom), 66

Sonja Bullaty 65 (bottom), 77 (bottom)

Ed Cooper 19 (top), 63, 74, 76 (bottom)

Kent and Donna Dannen 19 (bottom)

Richard Day 6, 29

Dembinsky Photo Associates
Dominique Braud 49 (bottom)
Willard Clay 14 (bottom)
Sharon Cummings 8-9, 13 (bottom), 34 (top)
Doug Locke 67, 72-73
Dreda Murphy 11
Alan G. Nelson 71 (bottom)
Ken Scott 10, 48
Stephen J. Shaluta 44

Jack A. Keller 68 (bottom)

Angelo Lomeo 49 (top), 65 (top)

Alan K. Mallams 64, 70

Patti McConville 20, 22, 31 (top), 34 (bottom), 36 (bottom), 37, 45, 56-57, 58, 61 (top), 78 (bottom)

New England Stock Photo
Roger Bickel 43, 68 (top)
Fred M. Dole 69
Michael Giannaccio 12
Thomas Mitchell 30 (top)
Effin Older 54 (top)
Margo Taussig Pinkertown 5, 16, 24-25
Darrin A. Schreder 71 (top)
Jim Schwabel 13 (top), 55
Kevin Shields 17
W.J. Talarowski 7
Jeremy Woodhouse 53 (top)

John Oram 26, 42 (bottom)

LeRoy G. Schultz 14 (top), 28 (top & bottom), 33, 42 (top), 50-51, 54 (bottom), 59, 60, 61 (bottom), 62 (top), 78 (top)

Scott T. Smith 46, 75 (top & bottom)

Stephen R. Swinburne 52 (top & bottom), 53 (bottom), 77 (top)

Terry Wild 35

David F. Wisse 32